THE UNIVERSE, EARTH & LIFE

I0494359

KALYANARAMAN B

An imprint of Notion Press

XpressPublishing
An imprint of Notion Press

Old No. 38, New No. 6
McNichols Road, Chetpet
Chennai - 600 031

First Published by Notion Press 2019
Copyright © Kalyanaraman B 2019
All Rights Reserved.

ISBN 978-1-64733-280-8

Contents

PREFACE

This book is a brief summary of my conclusions from the several articles and books I have read and presented in a simple language for the easy understanding of the subjects dealt with by the common man.

The author aged 73 years is a postgraduate in physical chemistry from the Institute of Science, Mumbai university.

He has worked in India and abroad and has travelled to many countries.

He is married and has two sons settled in USA & UK respectively.

The author would welcome comments and suggestions for improvements.

Kalyanaraman B

I

THE UNIVERSE

Everything known to humans can be reduced to at least anyone of the following four fundamental irreducible absolutes, that collectively constitutes the universe:

1.Space
2.Matter-Energy
3.Laws of Nature
4.Consciousness.

The observable material universe is an incessant evolving reality based on interactions among the four fundamental absolutes mentioned above leading to the unlimited flow of infinite events.

While Space, Laws of Nature and Consciousness are independent, invisible and intangible entities, only matter-energy manifests in different forms perceptible and visible to the humans.

Time is only an arbitrary earth-based standard of measure of the intervals between events invented by humans for their references & studies.

Scientific studies have identified the composition of the physical material universe as follows:

COMPOSITION OF THE MATERIAL UNIVERSE (MATTER-ENERGY)

One of the most surprising discoveries of the twentieth century is that the ordinary visible or baryonic matter comprising galaxies, stars & planets, made up of protons, neutrons and electrons bundled together into atoms and molecules, makes up less than 5% of the mass of the universe.

The rest of the universe appears to be made up of mysterious invisible substance called dark matter (25%) and an unknown force that repels gravity called as dark energy (70%).

Dark matter (25%) is invisible, neither reacts with baryonic matter nor interacts with other forms of radiations making it impossible to detect with current instruments. But scientists are confident it exists because of the gravitational effects it appears to have on galaxies and galaxy clusters.

Most scientists think that dark matter is composed of non-baryonic matter called **WIMP**(Weakly Interacting Massive Particles).

Dark energy (70%) discovered in 1990s is even more mysterious. Scientists now think that it is a repulsive force against gravity generated by quantum fluctuations in otherwise empty space and drives the accelerated expansion of the universe.

According to the well-established laws of thermodynamics, neither matter-energy can be created from nothing nor destroyed into nothing through any process known to humans by anyone, anywhere and at any time in the history of this universe but can only be interconverted or transformed from one form into another.

This means that the total amount of matter-energy in the universe is constant and can neither be increased nor decreased by any process.

If these scientific laws have eternal validity, the material universe as constituted above could only either be a self-existing reality on its own or must have evolved from an already self-existing (unknown) entity without a cause, creation, purpose or beginning.

QUANTUM MECHANICS

Classical physics deals with properties, nature and motions of macroscopic matter that we encounter every day and can see and touch without the need for additional tools.

Quantum physics or Quantum Mechanics deals with properties, nature, motions and interactions of microscopic & subatomic particles and other radiations we cannot see directly with our eyes.

As a result of scientific investigations during the last four centuries, all material objects have been found to consist of not only tiny, rigid material particles but also an intangible constituent of energy which may be set free from all association with matter when it travels as radiant energy or radiation.

These radiations manifest themselves in different natures such as the electromagnetic radiations, the Hertzian waves, the infrared radiations, the visible radiations, the ultraviolet radiations, the X radiations, the Alpha, Beta & the Gama called as nuclear radiations and finally coming to the lowermost wavelength or the highest frequency the cosmic radiations whose origin and source are as yet unknown to the mankind.

According to quantum theory all radiations takes place not continuously but discontinuously and discreetly as elementary multiples of a unit of energy called "quanta" which have both the properties of a particle (mass & velocity) and a wave (wavelength & frequency) until they are observed.

The very process & nature of observation of radiations makes it reveal its identity as a wave or a particle. This has been amply demonstrated and repeatedly verified by the famous double slit experiment in the early part of the twentieth century.

The particle and the wave aspects are not mutually exclusive, in the sense that reality whether matter or radiation, is made up of a subtle and an almost indefinable fusion of two antagonistic but complimentary factors, the continuous waves or discontinuous particles. It is a continuous discontinuity or a discontinuous continuity and hence not a simple but a complex unity.

According to Werner Heisenberg's uncertainty principle, both the position and the momentum of a particle in motion cannot be determined accurately at the same time. This is because the very act of measurement involves interaction with another particle of similar magnitude and dimension which inevitably disturbs its position and or speed at the precise moment of impact.

The laws of classical physics and quantum physics are not complimentary and where one applies the other does not. Scientists are still working towards establishing a unified theory of the laws of physics in general.

THE BIG BANG THEORY

In the late 1920s the American astronomer Edwin Hubble made a very interesting and important observation that distant stars and galaxies are receding from earth in all directions with velocities in proportion to their distance from earth. This observation has been confirmed by numerous experiments and repeated measurements. The implication of these findings is that the universe is expanding.

This led to the conclusion, based on the knowledge of developments in astrophysics, that all the currently observed matter and energy in the universe were initially condensed in a very small hot mass with infinite density and gravity called as a Singularity.

The present dominant scientific concept about the origin of this universe is the Big Bang theory, first proposed by Belgian Catholic priest Georges

Lemaitre in 1927, according to which approximately 13.8 billion years ago a huge explosion, known as the Big Bang, disintegrated the Singularity mentioned above and space started to expand spreading matter and energy in all directions.

All of space, matter-energy & laws of nature evolved at the same moment from the Big Bang expansion and Intervals of events from this moment are estimated and measured as time by humans through an arbitrary earth based standard.

However, science cannot reconstruct what happened before the Big Bang or how the infinitely dense singularity came into existence or why it started to disintegrate and expand at a point in time as all the known laws of science breakdown and cease to operate at the point of singularity.

Although the Big Bang theory is widely accepted by the scientists based on available demonstrable evidences , it probably will never be firmly established as we cannot travel back in time to verify and validate the variety of events predicted by the theory immediately within few seconds after the Big Bang expansion started.

Even if there is/was a GOD, prior to the Big Bang expansion, He has apparently ceased to manifest and/or interfere thereafter in the operations of this universe which are now governed and regulated only by the immutable eternal laws of nature.

Evidence for the Big Bang Expansion:

The Big Bang theory predicts the existence of cosmic background radiation with temperatures several degrees above absolute zero in deep space. This radiation was discovered in 1964 by Arno Penzias and Robert Wilson who later won the Nobel prize for this discovery. The Cosmic Microwave Background Explorer (COBE) satellite launched in 1991 confirmed that the background radiation field has exactly the spectrum predicted by the Big Bang theory.

By the Big Bang disintegration and expansion of the infinitely dense singularity energy is converted into a variety of subatomic matter and antimatter particles. But the conversion produced a little more matter than antimatter (science cannot explain why). Most antimatter have annihilated with matter producing energy. The remaining matter is scattered in all directions due to the expansion and begins to cool.

First quarks clump together into protons and neutrons and then protons capture electrons to form Hydrogen atoms.

The explosion of the Big Bang only formed Hydrogen and Helium.

Later, Gravity pulls the Hydrogen and Helium together to form stars and galaxies.

Heavier elements (Carbon, Oxygen, Nitrogen etc.) are formed inside stars through thermonuclear fusions.

Stars burned out and exploded and expelled the heavier elements that formed planets.

The ages of the universe, our galaxy, the solar system and earth can be estimated using variety of modern scientific methods.

Over the past few decades, measurement of the Hubble expansion has led to estimated ages for the universe between 7-20 billion years, with the most recent and best measurements within the range of 10-15 billion years.

WHEN WERE DARK MATTER & DARK ENERGY CREATED?

There is so much that is unknown about the dark matter and dark energy but there are great many things that can be said about them.

1. Dark energy affects the expansion of the universe, only becoming prominent and detectable about 6 – 9 billion years ago.

2. It appears to be the same in all directions.

3. It appears to have a constant energy density throughout time.

4. It appears to not clump or cluster or anticluster with matter and remain uniform throughout space.

We have no evidence that speaks one way or the other on dark energy's presence or absence for the first four billion years or so of the universe's history.

Dark matter on the other hand has shown its effects for the entire 13.8 billion years history of the universe.

Dark matter does not clump and cluster, and its effects can be seen in the formation of the earliest quasars, galaxies and clouds of gas.

Even prior to that, dark matter's gravitational effects show up in the earliest light from the universe: The cosmic microwave background or the leftover glow from the Big Bang.

The pattern of imperfections that the universe is made of cannot be explained without approximately 27% of dark matter.

Without knowing exactly what dark matter is, we cannot say with any certainty exactly when it may have arisen. But from the measurement of the large-scale structure of the universe we can be certain that dark matter arose in the early stages of the Big Bang, possibly at the very beginning of it all.

Dark energy may have been around the whole time, or it may have emerged only much latter. There is substantial basis for the idea that only when a complex structure forms, dark energy arises and becomes important in the universe.

EVOLUTION OF SOLAR SYSTEM

The formation and evolution of the solar system began with a gravitational collapse of a giant molecular cloud. Most of the collapsing masses collected in the centre and formed the sun, while the rest flattened into a protoplanetary disc out of which the planets, moons, asteroids and other solar system bodies formed.

The solar system is made up of a centrally located massive star Sun surrounded by eight planets revolving around it in different orbits. The planets with increasing distances from Sun are Mercury, Venus, Earth, Mars, Jupitar, Saturn, Neptune and Pluto.

Some of the planets including earth have moons revolving around them.

The best estimates of the earth's age are obtained by calculating the time required for development of the lead isotopes in earth's oldest lead ores. These estimates indicate 4.54 billion years as the age of earth and meteorites, and hence of the solar system.

The study of several forms of physical and geological evidences such as fossils, various types of rocks, volcanic eruptions, earthquakes, soil erosion, minerals, radioactive substances, meteorite hits etc have firmly confirmed that the earth is very old.

II

PLANET EARTH

Intelligent Design or Fortuitous Accident:

Since scientists agree that the earth has not existed eternally, but evolved through the Big Bang only, simple logic dictates that no middle position exists on the important issue of designed plan vs fortuitous accident.

Either a super conscious mind planned and designed our planet, or it all originated by a fortuitous accident.

At first it is hard to believe that GOD had anything to do with the Big Bang and from the location of the earth it looks and appears that it is nothing more than a meaningless speck of dust lost in the unlimited vastness of the universe.

However, the more Astronomy learns about the universe and Geology about Earth, the more scientists believe that the earth IS NOT JUST a meaningless speck but has been UNIQUELY DESGNED and PLACED.

Out of the millions and millions of planets in the universe the earth is the only planet known to have evolved a life supporting system.

Let us consider some amazing facts about the evolution of earth to resolve the issue of design vs evolution.

1.EARTH'S SURFACE TEMPERATURE

The most important factor affecting the earth's surface temperature is obviously the distance from the sun. The earth's orbit around the sun is approximately 93 million miles away from the sun.

If the earth's orbit were moved 1% closer to the sun, there will be drastic heating effect on the earth's surface and life as we know will cease to exist.

If the earth's orbit is moved 1% farther from the sun the reverse of the previous situation applies. The earth's surface will be too cold to support normal life.

It appears that the earth is just the proper distance from the sun to maintain the right range of surface temperature suitable for life and the many important geological processes!

2. EARTH'S TILT & ROTATION

The earth's axis of rotation is tilted 23.50 degrees relative to the perpendicular of the earth's plane of orbit. This tilt causes the four seasons.

If there were no tilt in the axis of rotation, we would have no seasons and the surface temperature at any point on earth would always remain the same. The equatorial region of the planet would be intolerably hot all year and the poles will remain ice cold.

Only the mid-latitudes will be comfortable for human habitation and suitable for cultivation. Only about one half of our presently farmable lands could grow crops.

If the earth had double the present tilt temperature extremes between seasons would be much more pronounced. Even mid latitudes would have unbearable heat in the summer and frigid cold in the winter. Life on most of the earth's surface would become intolerable.

The earth rotates once every 24 hrs producing day and night. If the earth rotated more slowly, we would have more extreme day and night temperatures. If the earth rotated more quickly the average surface temperature would go up.

Thus, we could hardly improve on the present arrangement of tilt and rotation, which seems to be planned for comfort and economy producing maximum farmable lands and pleasant seasons.

3. EARTH'S ATMOSPHERE

The earth's atmosphere is composed of four important gases.

The most abundant gas is Nitrogen which comprises about 78% of the atmosphere. Oxygen gas is the second most common ingredient with 21% Argon is third with slightly less than 1%. Fourth is carbon dioxide with 0.03%.

While Argon is inert, Nitrogen is relatively inactive and enters very few chemical reactions. It is indeed fortunate that nitrogen gas does not readily combine with oxygen; otherwise we could have an ocean full of nitric acid.

Oxygen is the most common reactive gas in our atmosphere and at 21% is the most abundant in earth out of all the planets known in the universe. More than trace amounts have not been discovered in the atmosphere of any other planet.

The present level of oxygen seems to be optimum.

If we had more oxygen, combustion will occur more energetically, rocks and metals would weather faster, and life would be adversely affected.

If oxygen were less abundant respiration would be more difficult and we would have a decreased quantity of ozone gas in the upper atmosphere which shields the earth's surface from deadly ultraviolet rays.

Carbon dioxide is only 0.03% of our atmosphere but seems to be at the optimum value.

If we had less carbon dioxide, the total mass of terrestrial and marine plants would decrease, providing less food for animals, the oceans would contain less bicarbonate, becoming more acidic, and the climate would become colder due to the increased transparency of the atmosphere to heat.

While an increase in atmospheric carbon dioxide would cause plants to flourish, there would be some unfortunate side effects. There will be rise in surface temperature, chemical weathering, rising bicarbonate levels in oceans leading to alkaline conditions unfavourable to life.

The analysis shows that our atmosphere has both the correct composition and density.

Is the present composition of earth's atmosphere a freak accident?

4. WATER & WORLD OCEAN

Density of WATER

Water is one of the very few substances known to humans whose density in its solid state is less than in its liquid state.

Consequently, ICE floats in water.

This very nature of water is very vital to LIFE and is extremely unique among the millions & millions of chemical compounds that exist in this universe.

The "scientific explanation" of this unique property goes into the crystal structure of ice and so on but cannot explain why this property is not found in any other chemical compound.

It is as if someone foresaw that without this property in ice, there won't be life and DESIGNED this property in ice so that life CAN exist.

Water is an extremely rare compound in space. A permanent reserve of liquid water is known to exist only on the earth estimated at some 340 million cubic miles of liquid water.

Water in liquid form has many unique chemical and physical properties which make it ideal as the primary component of life.

The solvent characteristic of water makes it possible for all essential nutrients needed by life to be dissolved and assimilated.

The fact that water is transparent to visible light makes it possible for marine algae to perform photosynthesis below ocean surface and for ocean animals to be able to see through water.

Obviously without water there cannot be life on earth.

5. EARTH'S CRUST

The continents which cover 29% of our planet's surface have a mean elevation of about 2,750 feet above sea level. The world ocean which covers 71% of the earth's surface has an average depth of some 12,500 feet!

If we were to scrap off the continents and place them in the deeper parts of the ocean to make an earth of common elevation, we would have an earth covered with approximately 8,000 feet of water!

There are two main reasons why continents remain elevated above the sea floor.

1. The continents are made up of rocks which are less dense than the rocks of the ocean bottom.

2. The continental crust is usually over twice as thick as the oceanic crust.

The difference in density and thickness between continental and oceanic crust is just about the right amount to maintain the present "freeboard" of the continents above the ocean bottom.

Study of meteorites has revealed that the elements iron and oxygen are about equal abundance on the average. From what is known about the density and structure of the earth, geologists suggest that iron is the commonest element in the bulk earth, being slightly more abundant than oxygen. However, when the crust of the earth is considered, geologists estimate that oxygen is about eight times more abundant than iron! Furthermore, the earth's crust has unusually large amounts of silicon and aluminium.

If we had larger amounts of iron and magnesium in the crust, oxygen from the atmosphere would be consumed to weather these elements and an oxygen rich atmosphere would be impossible. Our present crust, unlike other planets and meteorites, is already highly oxidised and therefore permits an oxidising atmosphere.

Are we very fortunate to have a highly oxidised earth crust to permit an oxidising atmosphere?

6. LAWS OF PHYSICS

There are FOUR FUNDAMENTAL FORCES in Physics:

1. Gravitational Force (Weakest)
2. Electromagnetic Force
3. Nuclear Weak Force (A repelling force that causes radioactivity)
4. Nuclear Strong Force (Strongest that holds the nucleus together)

All the four fundamental forces were discovered and measured EMPIRICALLY and INDEPENDENTLY. That means there is no obvious relationship between these four forces.

With the advent of powerful computers physicists studied the effects when the strength of some force is changed. The findings are amazing.

1.Changing the strength of the Gravitational Force.

The gravitational force pulls all matters together. The gravitational force is the primary force in the formation of stars.

The gravitational force is 10^{39} times weaker than the electromagnetic force

If the gravitational force is made slightly strong and only 10^{33} times weaker than the electromagnetic force, then stars will burn million times faster and life will not have time or opportunity to develop (Radiation will kill all life)

If the gravitational force is made slightly weak, stars will burn too slowly and will not produce enough energy to support life.

2. Changing the strength of the Electromagnetic force.

If the electromagnetic force is a little stronger, the positive nucleus will hold on so strongly to their electrons that chemical reactions cannot occur. No chemical reactions, no valence bonds, no water, no carbon chains, no life.

If the electromagnetic force is a little weaker, the nucleus cannot hold their electrons in orbit and there will be no atoms.

3. Changing the strength of the Nuclear Strong Force

The strong force determines HOW STRONG protons and neutrons will attract to each other.

If the strong force is just 0.3% stronger, protons and neutrons will attract so much that solitaire protons will be EXTREMELY RARE.

In other words, there will be NO HYDROGEN ATOMS which means NO WATER MOLECULES and NO LIFE.

If the strong force is 0.2% weaker, protons and neutrons will not "stick together" and heavy elements cannot be formed.

The universe will consist of only hydrogen atoms without any CARBON or OXYGEN atoms and again NO LIFE will be possible.

Summary:
The relative strengths of the four fundamental forces must be within very narrow margins to allow for:
The existence of atoms
Chemical Reactions
Energy production in stars, etc, etc.
That allows life to exist in this universe.

Question:
Why are the physical constants of the fundamental forces within very narrow limits?

If the universe is a random product, it is extremely unlikely that the physical constants will all fall within very narrow margins to allow for building blocks of life to exist.

Recent discoveries have established that our earth and solar system is unique, and it may even be the only place in the whole universe that can harbour life.

Some finely tuned qualities of our solar system to support life.

1.The Mass of our Galaxy.

Our galaxy is on top 1% or 2% of the most massive and luminous galaxies in the whole universe.

This allows stars to be more massive and produce heavy elements which are essential for life.

2.The Mass of our Sun

The speed of nuclear reaction is dependent on the mass of the sun.

The mass of our sun is optimal for photosynthesis: all colours are represented.

If the mass is too low, the sun will burn (too slow) red and produce little blue light.

If the mass is too high, the sun will burn (too fast) blue and produce a lot of harmful ultraviolet light.

3.Jupitar (Saturn & Neptune): Earth's Guardian Angel

The Giant Planet Jupitar (and in a lesser amount, Saturn & Neptune) is SHEILDING earth from many asteroids / comets by drawing these space debris to itself and away from hitting earth.

Jupitar is in fact "taking the hits" for earth.

Evidence of this is the collision of the comet Shoemaker- Levi 9 with Jupitar in July 1994.

4. Earth's Moon

Earth's moon has several extraordinary properties that are extremely rare.

a) Moon's Size

No planet has been found that has a moon that is THAT HUGE compared to the host planet's size.

b) Moon's Orbit

The eccentricity (flatness of the elliptical orbit) of the lunar orbit is very small – almost circular.

Moon & Earth's Tilt

The earth's tilt ranges between 22 degrees to 24.50 degrees over a 41,000-year cycle.

Smaller tilt means STABLE temperature & environment for life.

c) Moon EXACT matching SIZE and DISTANCE

1.The moon is 400 times smaller than the sun.

2. The moon is also 400 times closer than the sun.

d) Mystery of the Moon's Origin

How is Earth's moon created?

Due to the huge size of the moon and the CIRCULAR lunar orbit, it is ALMOST IMPOSSIBLE that the moon was CAPTURED by EARTH.

The "BEST explanation" so far is:

When Earth was in the forming stages, a Mars – size object collided with earth and some of the material spun into orbit and formed the moon.

The problems with this "BEST EXPLANATION" are:

1. The object cannot be too small or too large. It must be a bit smaller than the earth itself – size of MARS and MARS size objects flying around the universe is very rare.

2. The chance that the RIGHT SIZE object hitting the earth is EVEN RARER.

3. Computer simulation has shown that the RIGHT object must hit earth at a VERY PRECISE angle to create a moon.

If the angle is too steep, the object is absorbed into the earth and forms a larger planet.

If the angle is not steep enough, the object bounces into deep space.

Summary: The probability of a MARS size object hitting the earth and creating our moon is virtually zero.

Basically, we do not have a clue as to how our moon could be so big, has a circular orbit and happens to be there to stabilise earth's climate.

Without the moon the earth's climate would not be able to sustain normal life (except may be for some bacteria that can survive freezing cold and blistering heat).

CONCLUSION:

Two different conclusions can be drawn from the data presented above.

1. Fortuitous Accident: It takes a great deal of faith to conclude that a very large number of extremely improbable events could occur by chance or accident in the same location with very precise accuracy that would lead to the evolution of life as we see on earth.

It is almost akin to supposing that the best model of a BMW car came into existence from globs of steel hurled on a heap of sand in a deserted island through an improbable, if not an impossible, accident by chance.

2. The creationist, on the other hand, will recognise that the only rational deduction from the data is that the marvels of the earth owe their origin to the handiwork of a superhuman designing intelligence popularly referred to as GOD.

However, this would lead to infinite regression, who created the ultimate creator?

III

Life Evolution on Earth

The relative size of earth in this universe is like a microscopic fraction of a grain of sand out of all the seashores in the world. So infinitely enormous is the magnitude of the insignificance of earth in this universe.

It requires an amazing combination of extraordinarily unique properties and compositions of matter in the right proportions, incredibly benign physical environment and exquisitely precise scientific laws of nature for the evolution of life supporting system anywhere in the universe.

The statistical improbability of such an incomprehensible scientific reality purely by chance or accident as evidenced by the known presence of life only on earth in the entire universe suggests the involvement of a supernatural designing intelligence in the evolution of life supporting system for the different kinds of living beings with innumerable variations in their faculties..

Nevertheless, no universally demonstrable evidences are available to validate and conclusively establish the identity of such an entity and remains only in the speculative domain of abstract religious beliefs.

The solar system and earth were formed 4.5 billion years (Ga) ago and evidence suggests that first primitive life emerged around 3.70 Ga.

Biologists reason that all living organisms on earth must share a single last universal ancestor, because it would be extremely improbable that two or more separate lineages could have independently developed the many complex biochemical mechanisms common to all living organisms, by sheer coincidence.

Approximately 1 trillion species currently live on earth of which only 1.70 -1.80 million have been named and 1.60 million documented in a central database. The currently living species represent less than one percent of all species that have ever lived on earth.

Life on earth is based on carbon, water and a handful of other elements.

Carbon with a valency of four and an ability to form covalent bonds provides stable framework for complex chemicals and can be easily extracted from the environment especially from carbon dioxide.

There is no other chemical element whose properties are similar enough to carbon to be called an analogue. The other elements closer to carbon in periodic table like boron, nitrogen etc can all put together form only a very small fraction of the number of compounds that carbon can form.

Silicon has also a valency of four but cannot form many compounds like carbon because of its stereochemical structure.

Boron & nitrogen, respectively, has only one electron less and only one electron more than the carbon atom. This difference of one electron must ultimately account for the presence and absence of life on earth.

Nevertheless, life with organisms based on alternate biochemistry, unknown to humans, may be possible in other planets.

Fossil studies indicate that many organisms that lived long ago are extinct and the extinction of species is so common that an estimated 99% of the species that have ever lived on the earth no longer exist.

The basic evolution timeline of a 4.6 billion-year-old Earth includes the following:

1. About 3.5 – 3.8 billion years of simple CELLS (PROKARYOTES)
2. 3 billion years of PHOTOSYNTHESIS
3. 2 Billion years of COMPLEX CELLS (EUKARYOTES)
4. 1 billion years of MULTICELLULAR life.
5. 600 million years of simple ANIMALS
6. 570 million years of ANTHROPODS (INSECTS, SPIDERS etc)
7. 550 million years of COMPLEX ANIMALS
8. 500 million years of FISH & PROTO-AMPHIBIANS
9. 475 million years of LAND PLANTS
10. 400 million years of INSECTS & SEEDS
11. 360 million years of AMPHIBIANS
12. 300 million years of REPTILES
13. 200 million years of MAMMALS
14. 150 million years of BIRDS
15. 130 million years of FLOWERS

16. 65 million years since non-avian DINOSAURS died out
17. 2.5 million years since the appearance of HOMO
18. 200,000 years since the appearance of modern HUMANS
19. 25,000 years since NEANDERTHALS died out

The idea that, along with other life forms, modern day humans evolved from an ancient, common ancestor was first proposed by Robert Chambers in 1844 and taken up by Charles Darwin in 1871.

Since the time Charles Darwin's theory of biological evolution of living species was first proposed, several developments have taken place in molecular biology, genetic engineering and living sciences that have increased our understanding of how life could have evolved in the early earth.

Based on latest molecular genetic data, evolutionists currently favour the hypothesis that modern Homo sapiens, individuals very much like us are descendants of a single small archaic population first evolved in Africa which then migrated all over the world less than 200,000 years ago.

While the theory of evolution can trace back the historical origin of first life, our present scientific knowledge is still far too inadequate to explain the evolution process of first life from inanimate chemicals only but leaves it to an improbable accident of nature by chance.

The evolution of volitional consciousness from inanimate matter, as an integral faculty of only human beings, continues to remain as a profound scientific mystery.

Creationism, intelligent design, and other claims of supernatural intervention in the origin of the universe, life or species are not science because they are not universally testable by the methods of science. These claims subordinate observed data to statements made on authority, revelation or religious belief.

IV

HUMAN LIFE

We are all born as transient entities with a range bound duration of existence, and have no choice about our physical, physiological or intellectual faculties and material resources at the time of our birth, but must compete with billions of other entities, perhaps more or less generously endowed, and progress in life to achieve success and happiness only through judicious use of the faculties and resources within our command and control, in an environment where many other factors outside our realm of intelligence, knowledge, influence and power also determine the course and the results of our actions at all points of time.

To the extent that we can understand and appreciate the profound complexity of this inexorable and inevitable fundamental reality, serenity, grace and dignity will guide us to accept the irresistible consequences of our actions and the incessant flow of events in complete equanimity, with neither a smile nor a frown, that would lead us to a sublime blissful state of mind full of inner peace and abiding tranquillity.